WORLD'S BIGGEST ANIMALS

GIANT WETA MAMMOTH INSECT

NATALIE HUMPHREY

PowerKiDS press

Published in 2024 by The Rosen Publishing Group, Inc.
2544 Clinton Street, Buffalo, NY 14224

First Edition

Editor: Natalie Humphrey
Book Design: Tanya Dellaccio Keeney

Photo Credits: Cover, pp. 1, 13 Graham R Prentice/Shutterstock.com; p. 5 Sid Mosdell/Flickr.com; p. 7 https://commons.wikimedia.org/wiki/File:Wetapunga.jpg; p. 9 Pyty/Shutterstock.com; pp. 11, 15 Br3nda/Flickr.com; p. 17 Kim Howell/Shutterstock.com; p. 19 Debbie Oetgen/Shutterstock.com; p. 21 Viktor Hejna/Shutterstock.com.

Library of Congress Cataloging-in-Publication Data
Names: Humphrey, Natalie, author.
Title: Giant weta : mammoth insect / Natalie Humphrey.
Description: Buffalo, New York : PowerKids Press, [2024] | Series: World's biggest animals | Includes index.
Identifiers: LCCN 2022055468 | ISBN 9781642825558 (library binding) | ISBN 9781642825541 (paperback) | ISBN 9781642825565 (ebook)
Subjects: LCSH: Giant wetas–Juvenile literature. | Grasshoppers–Juvenile literature.
Classification: LCC QL508.A56 H86 2024 | DDC 595.7/26–dc23/eng/20221223
LC record available at https://lccn.loc.gov/2022055468

Manufactured in the United States of America

CPSIA Compliance Information: Batch #CSPK24. For Further Information contact Rosen Publishing at 1-800-237-9932.

CONTENTS

The Biggest Bug

What can be heavier than a small bird but looks like a cricket? The giant weta! There are 11 different **species** of giant wetas. The biggest species is called the Little Barrier Island giant weta, or the wetapunga!

How Big Is a Weta?

Giant wetas are long and very heavy for bugs. A full-grown giant weta is usually 3 inches (7.6 cm) long and weighs between 0.3 and 1.2 ounces (9 and 35 g). But the heaviest wetapunga ever recorded was 2.5 ounces (71 g) and four inches (10 cm) long!

Where Are They From?

The giant weta only lives in New Zealand. Scientists studying the weta say that this bug has lived in New Zealand for more than 65 million years. That means giant wetas were around at the same time as the dinosaurs!

NEW ZEALAND

Awake at Night

Giant wetas are **nocturnal**. During the day, they sleep in trees, shrubs, or dead plants. At night, they move around to find food. Giant wetas usually stay off the ground but will come down to **mate**.

What's That Sound?

Wetas have a weird way of hearing. Their ears are on their knees! Giant wetas have holes on the inside of their knees. They use these holes to hear. When a giant weta wants to hear something in another direction, it moves its legs that way.

EAR HOLE

Plants for Dinner

Giant wetas usually eat plants. They eat fresh and dead leaves. They eat almost any plant they can find around their home. If they're really hungry and they can't find other food, they may eat smaller bugs! But that's not very common.

Crawling Around

Even though giant wetas look like crickets, they don't usually jump. Instead, they have hooks on their feet that help them walk and climb trees. If they face a **predator**, giant wetas will fight! They use their spiked back legs to fight off threats, or dangers.

Lots of Eggs

Female giant wetas lay a lot of eggs. One weta female can lay between 100 and 300 eggs! The female uses a special **organ** called an **ovipositor** to lay her eggs deep in the ground. Baby wetas, or nymphs, are very small and look just like adults.

OVIPOSITOR

Save the Wetas!

Giant wetas are **endangered**. But some people are helping make sure wetas will be around for years to come. One team of people from the Auckland Zoo in New Zealand has **released** nearly 4,000 giant wetas into the wild.

GLOSSARY

endangered: In danger of dying out.

mate: To come together to make babies.

nocturnal: Active at night.

organ: A part of an animal's body that does a specific job.

ovipositor: A long tube some animals use to lay eggs.

predator: An animal that hunts other animals for food.

release: To set something free.

species: A group of plants or animals that are all the same kind.

FOR MORE INFORMATION

BOOKS

Becker, Trudy. *Weta*. Lake Elmo, MN: Focus Readers, 2023.

Levy, Janey. *Giant Wetas Shock!* Buffalo, NY: Gareth Stevens Publishing, 2018.

WEBSITES

Auckland Zoo
www.aucklandzoo.co.nz/animals/weta-punga
Watch videos and learn more about giant wetas from some of the people trying to save them.

AZ Animals
a-z-animals.com/animals/giant-weta/
Learn more fun facts about giant wetas and how they live in the wild.

INDEX